SOMETHING ABOUT

X RAYS

For Everybody.

BY EDWARD TREVERT.

ILLUSTRATED.

LYNN, MASS.
BUBIER PUBLISHING CO.,
1896.

ORIGINALLY
COPYRIGHTED BY
BUBIER PUBLISHING
LYNN, MASS.
1896

REPRINTED BY
MEDICAL PHYSICS PUBLISHING CORPORATION
MADISON, WI
(A non-profit organization)
1988

FOREWORD.

X-rays were discovered in November 1895 by Professor W. C. Roentgen at the University of Wurzberg in Germany, and the medical applications of x-rays were immediately recognized in all advanced countries. In the United States, the first medical x-ray was taken February 3, 1896 by Professor Edwin B. Frost of Dartmouth University. He took an x-ray of a broken bone for a patient of his brother, Dr. G. D. Frost.

The wonders of x-rays were often discussed in the popular press. A book about x-rays, written for the public, was published in the summer of 1896, only about six months after that first x-ray was taken. This is a replica of that book. It is appropriate that it appear as we approach the centennial of the discovery of x-rays.

The copy from which this reprint was made was first owned by Professor C.W. Freeze. In the 1920s, he gave the book to one of his students, Albert E. Kidd, who later taught physics at Sam Houston University in Huntsville, Texas. Professor Kidd gave the book to one of his students, Leroy J. Humphries, in 1966. Dr. Humphries plans to donate the book to an appropriate library such as the American Institute of Physics' Neils Bohr Library at the Center for the History of Physics.

Funds for the reprinting of this book were provided by Dr. Humphries through the CNMC Company. At his request, one half of the net proceeds from sales will be donated to help fund the Robert J. Shalek Premasters and Predoctoral Fellowships in Clinical Physics at The University of Texas M.D. Anderson Cancer Center in Houston, Texas. The other half of the proceeds will be donated to Medical Physics Publishing Corporation, a non-profit organization devoted to promoting science education.

JOHN R. CAMERON, PH.D.
Professor Emeritus, University of Wisconsin-Madison
Departments of Medical Physics and Radiology,
September 1, 1988

PREFACE.

No scientific incident for many years has interested the general reader so much as Prof. Röntgen's discovery of the X Rays and their mysterious action upon the photographic plate and various fluorescent substances. The layman as well as the professional has been experimenting and trying to obtain knowledge of the source, action and effect of these rays.

Considering the great interest taken in this subject by almost everybody, I have thought it prudent to compile this little book which is intended for the general reader; in no way is it to be considered technical. Some of the articles are compiled from the leading electrical journals of the day, and I am particularly indebted to BUBIER'S POPULAR ELECTRICIAN, The *Electrical World*, The *Electrical Engineer*, *The Electrical Review*, *Electricity* and The *Western Electrician* and to The New York *World*, The *Boston Post* and The *Scientific American*, for such articles. I trust that my efforts may be of sufficient value to interest the reader.

EDWARD TREVERT.

LYNN, MASS., JUNE 10, 1896.

PROF. RÖNTGEN.

*"Wilhelm Conrad Röntgen was born in 1845 in Holland. He graduated at the University of Zurich, taking his doctor's degree at the age of twenty-five. At this university he was the favorite disciple of Prof. Kundt. When the latter left Zurich for Würzburg, Röntgen went with him, and the two next received appointments in Strasburg University as professor and assistant respectively. This was in 1873. In 1875 he held the chair of mathematics and physics at the Agricultural Academy of Hohenheim in the kingdom of Württemberg.

Hohenheim is a hamlet some four miles south-southeast of Stuttgart, little known except for its school of agriculture. He returned a year later to Strasburg, and in 1879 was professor in and director of the University and Institute of Physics in the old university town of Giessen, a city rendered illustrious before this time by the labors of the great Liebig. In 1888 he returned to his old college at Würzburg, where he now holds his professorship."

We are indebted to *L'Illustration* for the portrait of Prof. Röntgen which we reproduce on opposite page.

---

*Scientific American.

PROF. RÖNTGEN.

CONTENTS.

CHAPTER I.

The Intensity Coil and the Crookes Tube.

The intensity coil is necessary in the production of the X Rays, where a battery or a dynamo is used to furnish the primary current. It consists of two coils, a primary and secondary. To these are added, for the purpose of intensifying their action, a magnetic core consisting generallyof a bundle of iron wires. An alternating or pulsating current is passed through the primary coil and produces (by induction) a current likewise pulsating in the secondary coil, although the two coils are entirely unconnected. In order to raise the voltage of the secondary coil it is made of a finer wire than the primary and consists of many more turns. To give the reader a clearer idea of the intensity or induction coil we will give directions for making one. Proceed as follows:

The coils are to be wound upon a tube built up of cartridge paper, wrapped around a cylindrical stick $\frac{7}{8}$-inch in diameter. The tube is to be 8 inches long and the paper to be wound on until it is $\frac{1}{16}$-inch thick, making the tube 1 inch outside diameter. Each turn, or layer of paper is to be glued down to the preceding one, so that when dry you will have a good stiff foundation on which to begin work.

The ends of the reel upon which the wire is wound are to be made of well-seasoned hard

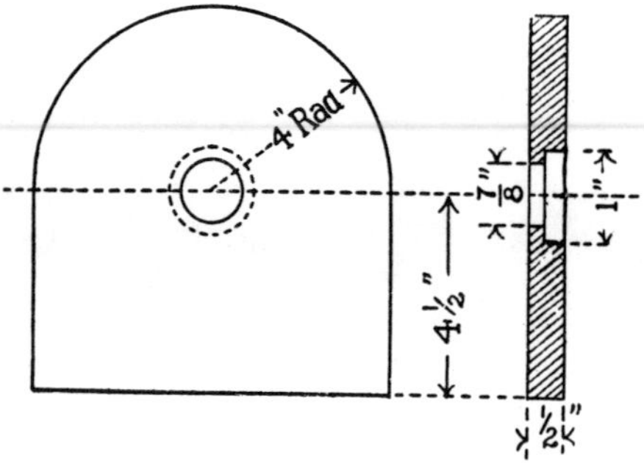

FIGURE I.

wood, thoroughly boiled in paraffin, of the shape and dimension indicated in the following sketch. (See Fig. 1).

These pieces are placed upon the ends of the paper tube and glued to it solidly.

Two holes are to be drilled in one of the end pieces, just below the tube, for the purpose of introducing the primary wire.

It will be well, if possible, to do the winding in a lathe, and for that purpose place the reel upon a snugly fitting wooden mandrel, with some sort of a dog to keep it from slipping.

Then, having pushed a short length, say a foot, of the wire with which you are going to wind your primary, through one of the holes in the end of the reel, start your lathe to revolving slowly and feed the wire on evenly, taking care to have it tight and to keep the insulation unbroken.

Use No. 16 cotton covered copper wire on the primary and wind on two layers, bringing

FIGURE 2.

the finishing end out through the other hole in the end of the reel.

Give the coil several coats of thin shellac, allowing it to dry thoroughly between each application, and then insulate it by winding on four layers of cartridge paper smoothly and glueing them down as directed in making the tube.

The paper should be cut say half an inch broader than the distance between the inside faces of the reel, and $\frac{1}{4}$ inch slits cut along the edge at short intervals, so that when placed upon the coil these slits allow the paper to be turned up at the edges and pasted against the ends of the reel. (See Fig. 2.)

The paper insulation should be made to fit as close to the ends of the reel as possible and whatever there may be in the way of interstices left between them should be carefully filled with as much shellac as they will soak up. It is also a good plan to fillet the corners with a mixture of resin and beeswax, five parts of the former to one of the latter.

The secondary wire is No. 36 silk covered.

On account of its liability to injury, the starting wire should not be brought outside the coil, but soldered to a piece of No. 18 gutta-percha covered wire, led in through a hole in the opposite end of the reel to that from which the primary wires issue. It is next to impossible to repair a break on the inside of the coil after it is once wound, and the greatest care should be taken while the winding is going on to make sure that everything is right before leaving it. On this account the gutta-percha covered lead wire should be wedged firmly into its exit hole, so that it cannot work around and break the joint inside.

Wind on the fine wire in the same way that

you did the coarse, evenly and smoothly, and when one layer is finished it should be thoroughly soaked in thin shellac and then covered with three thicknesses of thin writing paper, laid on with shellac. This paper should be made as directed above, a little wider than the distance between the inside faces of the ends of the reel and slit to fit, and the interstices well filled with shellac before going on with the next layer.

This is wound on and insulated in the same manner as the first, and the process continued until the coil is about 3½ inches in diameter. The finishing end should then be soldered to another piece of gutta-percha wire passing out of the opposite end of the reel to that from which the starting end issues, and the coil then covered with eight or ten layers of writing paper and wound over with silk ribbon or cord.

The danger of breaking is so great with this fine wire that each layer should be tested after being wound, by connecting the starting end of the coil to one pole of a magneto and the other pole to a spot on the wire, which you can bare for the moment and then cover up again when you resume winding. Be sure that your magneto will ring through the coil before you cover up the layer last wound.

Another method of testing, should you happen to be without a magneto, is to have a tolerably strong battery and connect one of its poles to the tine of a file and the other to the starting end, then with a piece of wire which touches a bare spot on the wire you are winding on, the roughened part of the file, if there is a good connection it will be shown by numerous sparks at the point of contact.

After winding, the coil may be mounted on its base.

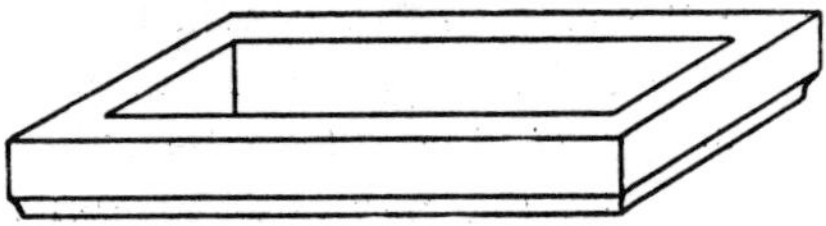

FIGURE 3.

This base is made of dry, well-seasoned wood, 8 inches wide, 14 inches long and 2 inches high. (See Fig. 3.)

The bottom is to be hollow and the edges and top to be $\frac{3}{4}$ of an inch thick.

The core for the coil is to be made of iron wires, about No. 12 or No. 14, cut to the same length as the reel and bound together by fine brass wires into a cylindrical bundle. Its outside diameter should be such that it will just slip inside of the tube upon which the coil is wound.

One end of this tube is to be closed by a wooden plug projecting into the tube $\frac{1}{4}$ of an inch, and this, of course, will cause the core to project a like distance from the other end of the tube.

The make and break device is to be mounted in front of this projecting end and is made according to the accompanying sketch. (See Fig. 4).

It consists of a piece of $\frac{3}{8}$ inch round iron, $\frac{1}{4}$ of an inch long and soldered to the end of a

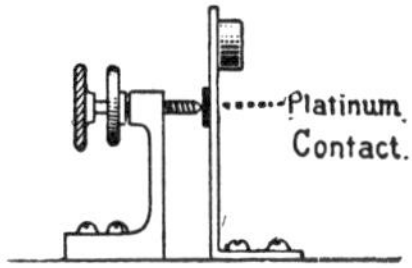

FIGURE 4.

piece of spring brass, whose other end is bent to a right angle and screwed to the base. To this spring is also soldered a small contact piece, preferably of platinum, but silver may be made to answer.

Opposite this contact piece and supported by a brass standard, through which it passes, is the contact screw, whose point is tipped with platinum. This screw has a check nut on it to prevent its working loose when the vibrator is in action.

The iron armature on the end of the vibrator must, of course, be opposite the end of the core and about $\frac{1}{4}$ of an inch away from it. After it is in place the spring can be bent backwards or forwards until the best adjustment is found.

A commutator will be found a very useful addition to the coil, for, while it is not absolutely essential it will prevent the unequal heating of the electrodes of the vacuum tubes used in some

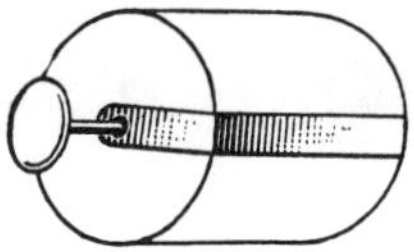

FIGURE 5.

FIGURE 6.

experiments, and will also serve as a key to open and close the primary circuit.

It consists of a cylinder of hard wood or vulcanite, 1 inch long and 1 inch in diameter, to which are affixed two brass strips $\frac{1}{4}$ of an inch wide. These strips are L shaped and are set opposite each other, with their edges flush with the surface of the cylinder. See Figures 5 and 6.

At the centre of each end of the cylinder is soldered a piece of brass wire to serve as an axis upon which the cylinder turns. The cylinder is mounted upon two supports made of sheet brass

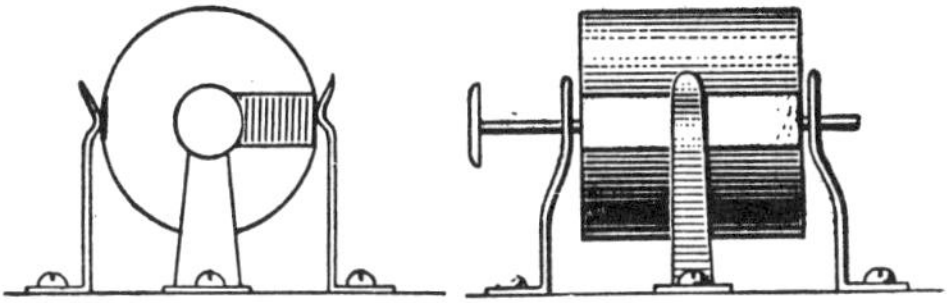

FIGURE 7.

and on each side of it is a spring brass contact piece. See Fig. 7.

The discharging apparatus next claims our attention.

The supports are to be made of heavy glass

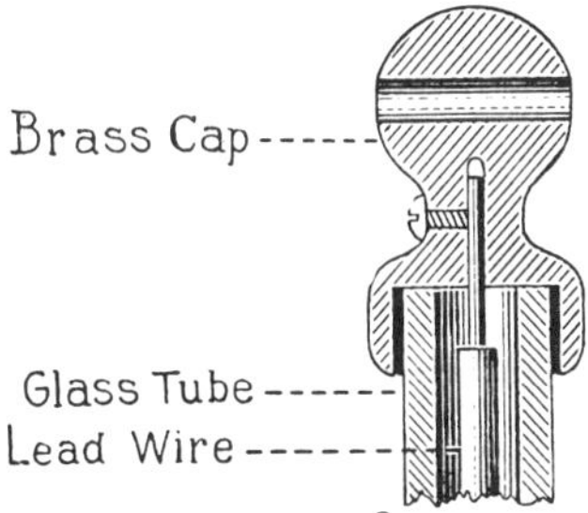

FIGURE 8.

tubing, such as is used in the water gauges on steam boilers. They are to be firmly set in the holes in the base, a little to one side of the central axis of the coil and to project about 4 inches above the top of the coil.

On the top of each tube a brass cap is fastened by means of sealing wax, as shown in Figs. 8 and 9.

The leading wires from the ends of the

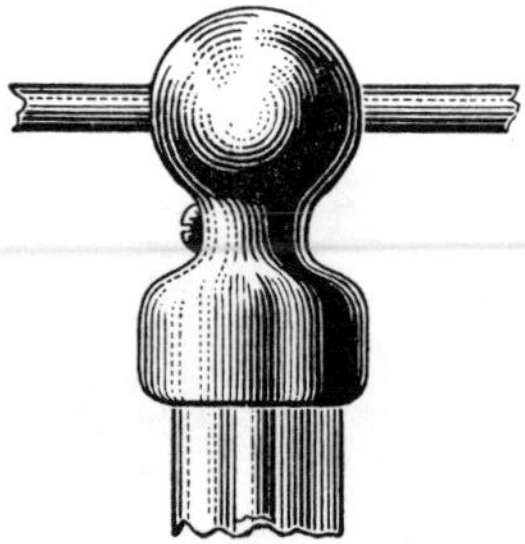

FIGURE 9.

secondary coils, which should both be gutta-percha covered, are to be taken by the nearest paths to the tubes, to the inside of them, through holes drilled for the purpose, and then

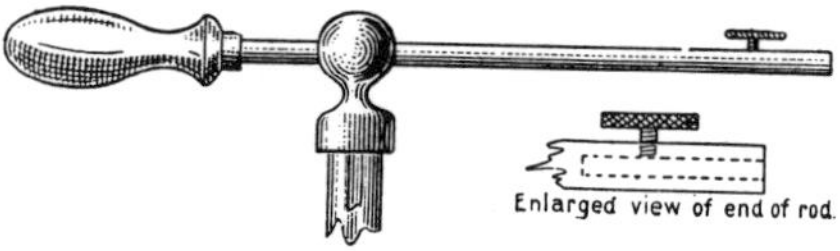

FIGURE 10.

up to the caps and fastened to them by screws, as shown in the sketches.

Holes may be drilled in glass with ordinary drills, whose points are kept wet with turpentine.

A discharging rod must be made for each cap piece. It must fit the hole in the cap snugly but not tightly, and be long enough to reach half way from one pillar to the other, and have room left on the other end for a handle.

The inboard end of each rod must be drilled about an inch deep and have a small screw with a milled head, with which to clamp anything which it may be desirable to attach to the rod. (See Fig. 10.)

The condenser for the coil may go into the

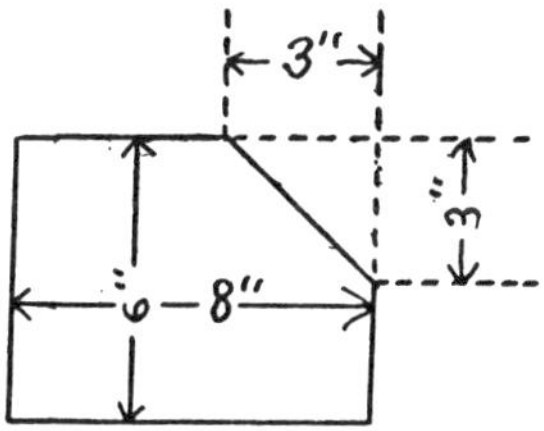

FIGURE 11.

hollow space in the base or may be placed in a separate box, but the first is preferable, as the instrument is in this way self-contained.

Prepare 50 sheets of tin foil 6x8 inches and cut off one corner of each, as shown in Fig. 11. You will also want 60 sheets of thin writing paper, cut according to Figure 12.

The cutting of both tin-foil and paper can, of

course, be done by hand, but it is better to fold up a number of large sheets to about the right size and take them to a bookbinder to trim to the shape you want.

Each paper must be examined carefully in a darkened room, before a gas jet or flame, for pin holes, and if any are found the piece having them must be rejected.

The paper is to be paraffined by dipping it

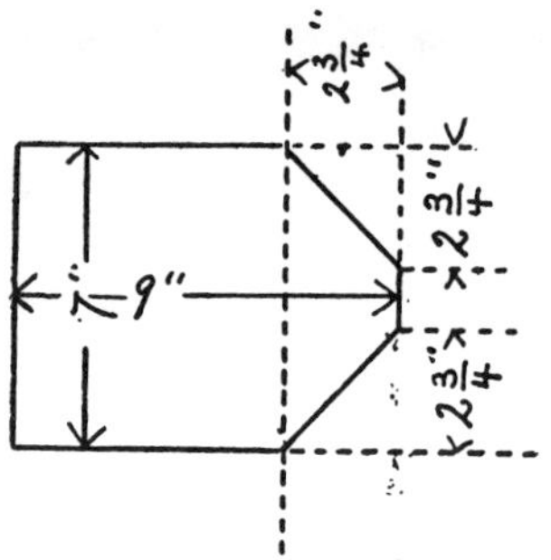

FIGURE 12.

in the melted wax. The melting must be carefully done so as not to cause the wax to boil or burn. After dipping a sheet draw it out and if the paraffin is hot enough the bubbles will run off. It must then be hung, by wooden clips, on a line already stretched for that purpose and the wax allowed to harden. When all the papers have been prepared we are ready to build the condenser. A compartment should be made in the base for the coil, just large enough to contain the condenser, that is, 9x7 inches.

Place five sheets of the waxed paper in this receptacle and then lay on them a sheet of tin-foil, with its unclipped corner projecting over one of the clipped corners of the paper. (See Fig. 13.) Then lay on a sheet of paper and on top of that another sheet of tin-foil, this time with its unclipped corner over the *other* clipped

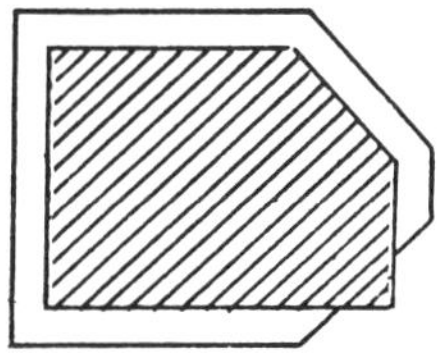

FIGURE 13.

corner of the paper. Then follows another sheet of paper and then one of tin-foil, laid on like the first, and so on, alternating the position of the unclipped corner with each sheet of tin-foil. When the last piece of tin-foil is in place put on the remaining sheets of paper and cover the whole with a small board of the same size as the compartment in which the condenser is placed. On the side of the board cover which turns towards the condenser glue two blocks, one over each of the uncovered corners of the tin-foil, and of such thickness that when the cover is in place they will press the sheets of tin-foil into compact bundles. (See Fig. 14).

Then through the side of the box holding the

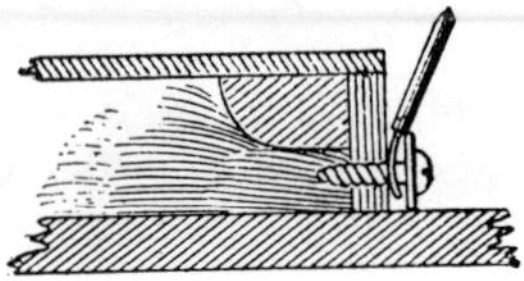

FIGURE 14.

condenser screw two short round-head wood screws far enough in so that their points shall penetrate the foil and to these screws attach the wires to the condenser. The connections may now be made according to Fig. 15.

Of course the wires are to be run beneath the base board as much as possible and wherever they cross one another they should be carefully insulated.

Fig. 16 is an elevation of the coil complete.

For battery power use three or four large Grenet cells, which are to be connected to the binding posts.

When the battery is connected and the commutator turned so that the brushes are touching the metal strips start the vibrator to working and adjust the contact screw until the motion is uniform and smooth. Then bring the ends of

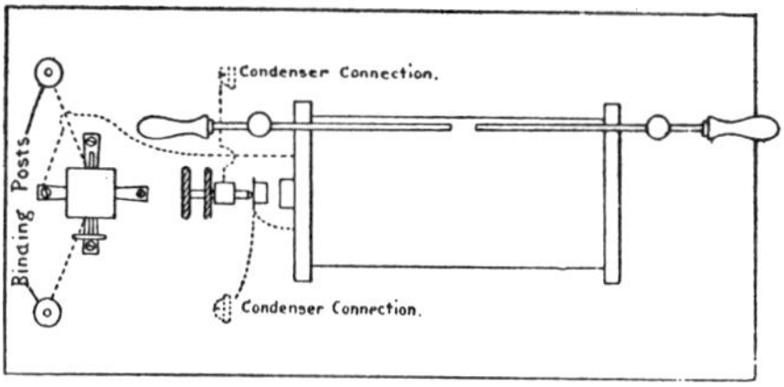

FIGURE 15.

the discharging rods (which have previously had short pieces of wire inserted in their ends for discharge points) together, when a shower of sparks will be observed to jump across the

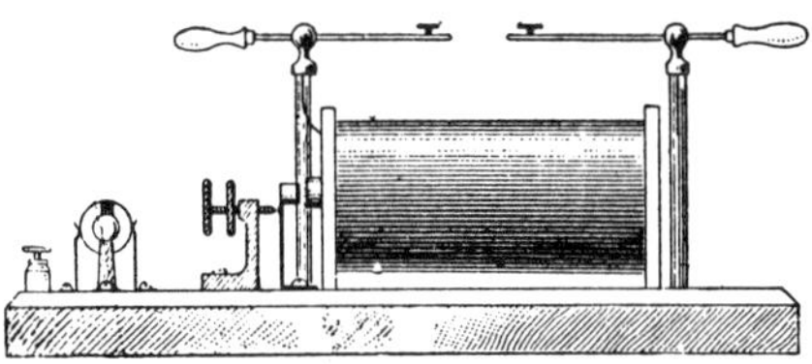

FIGURE 16.

space between them. The closer the points are together the more intense the sparks.

To get real good results in X Ray pictures,

you must have a coil that will at least give a spark from 1½ to 10 inches, to get good results with the fluorscope, from 4 to 10 inches. A Ruhmkorff coil is the best; this can be bought at most any electrical supply store and is made in different sizes.

The Crookes Tube.

The Crookes tube is another necessary adjunct to the production of X Rays. It consists of a thin glass globe or bulb into which is fused through the glass two platinum electrodes,

FIGURE 17A.

one for the anode and one for the cathode. The anode is the point where the current enters the tube. . The cathode is the point where the current leaves. Hence, either electrode of the tube, may be made the cathode, by simply reversing the direction of the current. A good tube not more than two inches in diameter and with the electrodes inside, not over an inch apart, with a coil giving a spark one and one-half inches, will make fairly good shadow pictures, if the exposure is long enough. If a

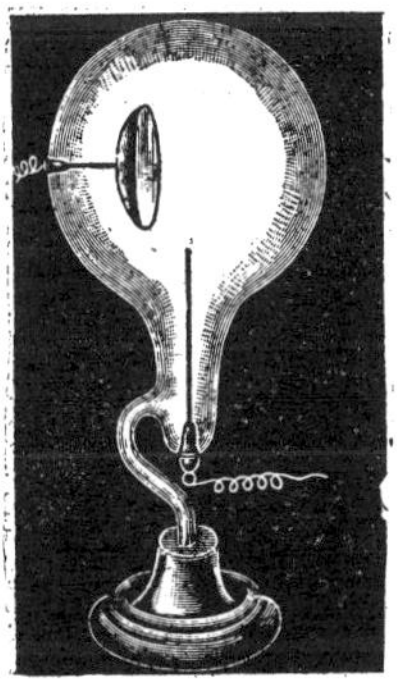

FIGURE 17B.

larger tube is used the coil should give a spark from three to ten inches. By this we mean that the secondary current should jump in the air the distance between the electrodes of the coil the above number of inches. If there is a blue glow that shows in a tube it will not produce X Rays. What light there is to be seen should be a green color. The glass bulb is exhausted of nearly all air and gases making nearly a vacuum. This vacuum differs from that in an incandescent lamp or a Geissler tube, viz.: It is so complete that only about one-millionth of the original quantity of atmosphere remains in the Crookes tube, while in the lamp or Geissler tube contains one thousandth or more. When a high potential intermittent or alternating discharge from the induction coil is sent through the Crookes tube there appears in the tube a fluorescent light of a green color. These rays are particularly strong around the cathode and are called cathode rays. The X Rays are the invisible rays which stream in straight lines from this point.

Prof. Crookes was the inventor of this tube, hence the name. They are made in various shapes; two forms are shown in Figs. 17A and 17B.

*A Proposed Standard Tube For Producing Roentgen Rays.

The observations which have up to this time been made as to the source of Röntgen rays point to the fact that the cathode rays must

*Elihu Thomson in the *Electrical World.*

strike some surface which then becomes the Röntgen ray source. My own observations make me think that if the bombarded surface is of such a dense metal as platinum, which does not fluoresce, the rays will be more energetic. Bearing in mind these observations, and the fact that it is desirable to possess Crookes tubes which are of standard make, and adapted for use with high frequency apparatus, Ruhmkorff coils with oscillatory or uni-directional discharge, or with Wimshurst or Holtz machines, I have proposed as a standard tube one which is represented in the accompanying illustration. See Fig. 18.

The structure consists of a bulb, at opposite sides of which are mounted concave aluminum discs supported on wires sealed through the glass. The foci of these concave electrodes coincide nearly at the centre of the bulb, at which point is located a V-shaped sheet of platinum, iron, a small piece of iridium, or other piece of metal. This could be supported in any way desired, but it may be suggested, as a convenience, that it be mounted on a third or middle terminal, so that it can be made an anode. In using this tube with high-frequency currents the two cup-shaped terminals are attached to the source of electric currents,

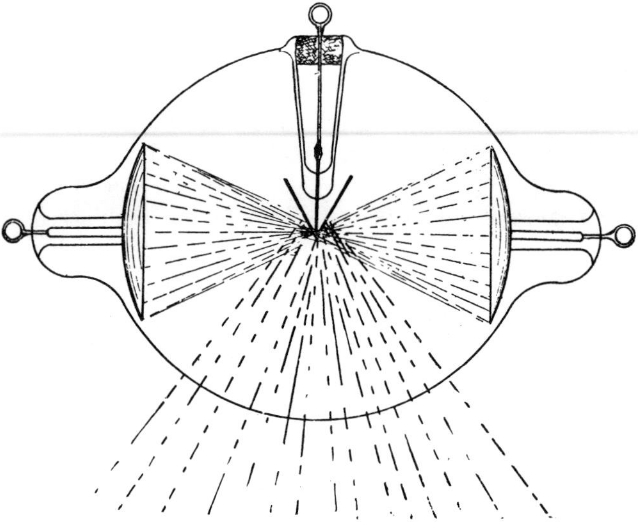

FIGURE 18.

which alternately become cathode during the discharges and cause the Röntgen rays to be emitted from the apex of the platinum sheet or from the surface of the piece of metal placed at the common focus. With oscillatory currents from a Ruhmkorff coil the same connections may be used, and even when the discharges are uni-directional the connection of either cup-shaped terminal as a cathode makes the tube effective, but in this case the middle terminal may be made the anode and the other or cup-shaped terminal a double cathode, and thus give the greatest amount of cathode surface with but a short distance to be traversed by the discharges, while the sharpness of the focus, and, consequently, the definition obtainable with the Röntgen rays remains unimpaired. If the vacuum in such a tube is too low, it is also possible to raise it quickly by constituting the middle terminal a cathode for a time, in which case metal would be carried therefrom and deposited on the side of the bulb in a very fine state of subdivision and quickly absorb the residual gas. If during the working of the tube the vacuum becomes too high it can sometimes be reduced by simply heating the tube to a high temperature.

CHAPTER II.

Experiments With X Rays.

Few events have agitated the scientific world into such active experimenting, as Prof. Röntgen's discovery of "X Rays" and their effect in shadow photography. By the aid of this new form of radiation we are enabled to photograph objects concealed in a box, a book, a leather case, etc., and even to lay bare the skeleton of a living or dead animal. Its application to surgery will be of great value, as by aid of these rays the surgeon may locate the exact position of a tumor, bullet, fracture, etc. Its value to chemistry, metallurgy, and other branches of science is at present inestimable.

The apparatus necessary to produce these rays is a Dynamo, Battery, or a Static machine, an Induction Coil and a Crookes tube or its equivalent, for example, an incandescent lamp with its filament broken and electrodes properly placed, and the bulb exhausted to the proper vacuum. The photographic plates used are the ordinary quick dry plates. The writer prefers "Seed."

With proper care and the necessary apparatus even an amateur may meet with wonderful success.

For the amateur a storage or Grenet battery is preferable. Either of these batteries may be

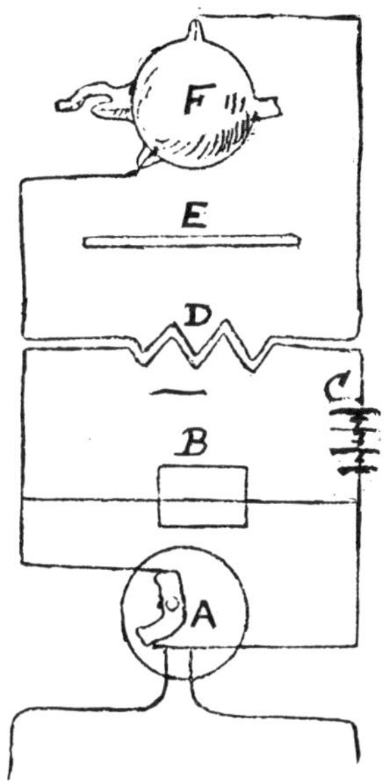

FIGURE 19.

obtained at an electrical supply store. For large coils of course a dynamo current is best, as for example, a coilgiving a spark of ten inches requires a current of 10 to 12 amperes to run it. In using an influence machine like a Holtz or Töpler, it has the disadvantage of reversing the current and spoiling the shadow picture.

Set up the apparatus as shown in the following diagram, Fig. 19. A is the current breaker,

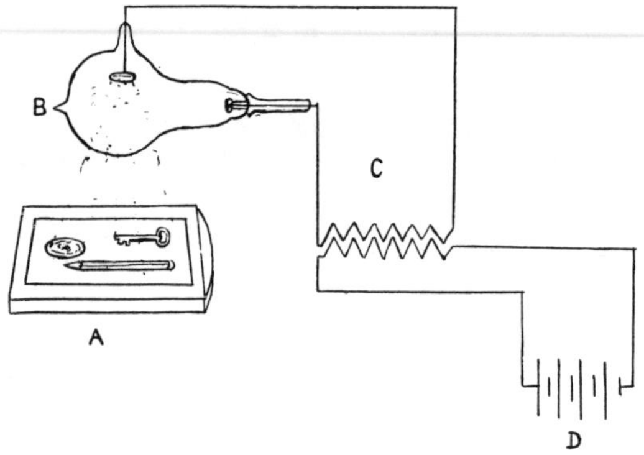

FIGURE 20.

which should run about 800 revolutions per minute. B is the condenser, C is the storage battery, D is the induction coil, E is the plate holder which contains a sensitised plate placed about 3 or 4 inches from F, the Crookes tube.

A diagram of a simpler apparatus is given in Fig. 20. A is the plate holder with objects lying upon it. B is the Crookes tube, C the high potential induction coil, D the storage battery consisting of at least 4 cells.

The object of which you wish to get a shadow picture must be placed between the Crookes tube and the plate holder.

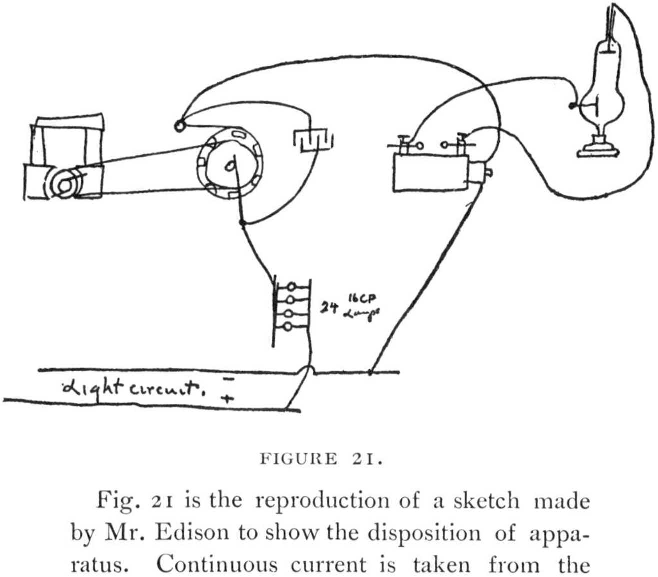

FIGURE 21.

Fig. 21 is the reproduction of a sketch made by Mr. Edison to show the disposition of apparatus. Continuous current is taken from the lighting mains, and a disruptive current produced by means of a disc with alternate conducting and insulating surfaces on its circumference, the disc being driven by the motor on the left.

Now for a word or two about the character of these rays. Prof. Röntgen says, "*If we pass the discharge from a large Ruhmkorff coil through a Hittorf or a sufficiently exhausted Lenard, Crookes or similar apparatus, and cover the tube with a somewhat closely-fitting mantle of thin black cardboard, we observe that in a completely darkened room that a paper screen washed with barium-platino-cyanide, lights up brilliantly and fluoresces equally well whether the treated side or the other be turned toward the discharge tube. Fluorescence is still observable two meters away from the apparatus. It is easy to convince one's self that the cause of the fluorescence is the discharge apparatus and nothing else.

The most striking feature of this phenomenon is that an influence (*Agens*) capable of exciting brilliant fluorescence is able to pass through the black cardboard cover, which transmits none of the ultra-violet rays of the sun or of the electric arc, and one immediately inquires whether other bodies possess this property. It is soon discovered that all bodies are transparent to this influence, but in various degrees. A few examples will suffice. Paper is very transpa-

---

* Abstract of Preliminary Communication to the Wurzburg Physico-Medical Society. Published by Messrs. Stahel of Wurzburg, who will shortly issue an English edition.

rent; the fluorescent screen held behind a bound volume of 1,000 pages still lighted up brightly; the printers' ink offered no perceptible obstacle.

Fluorescence was also noted behind two packs of cards; a few cards held between apparatus and screen made no perceptible difference. A single sheet of tin foil is scarcely noticeable; only after several layers have been laid on top of each other is a shadow clearly visible on the screen. Thick blocks of wood are also transparent; fir planks 2cm. to 3cm. thick are but very slightly opaque. A film of aluminum about 15mm. thick weakens the effect very considerably, though it does not entirely destroy the fluorescence. Several centimeters of vulcanized india-rubber let the rays through. Glass plates of the same thickness behave in a different way, according as they contain lead (flint glass) or not; the former are much less transparent than the latter. If the hand is held between the discharge tube and the screen, the dark shadow of the bones is visible within the slightly dark shadow of the hand. Water, bi-sulphide of carbon, and various other liquids behave in this respect as if they were very transparent. I was not able to determine whether water was more transparent than air. Behind plates of copper, silver, lead, gold, platinum, fluorescence is still clearly visible, but only when the plates are not too thick. Platinum 0.2mm thick is transparent; silver and copper sheets may be decidedly thicker. Lead 1.5mm. thick is as good as opaque, and was on this account often made use of. A wooden rod of 20x20mm. cross section, painted white, with lead paint on one side, behaves in a peculiar manner. When it is interposed between apparatus and screen it has almost no effect, when the X rays go through the rod parallel to the painted side, but it throws a dark shadow if the rays have to traverse the paint. Very similar to the metals themselves are their salts, whether solid or in solution.

3. These experimental results and others lead to the conclusion that the transparency of different substances of the same thickness is mainly conditioned by their density; no other property is in the least comparable with this.

The following experiments, however, show that density is not altogether alone in its influence. I experimented on the transparency of nearly the same thickness of glass, aluminum, calcspar and quartz. The density of these substances is nearly the same, and yet it was quite evident that the spar was decidedly less transparent than the other bodies, which were very much like each other in their behavior. I have not observed calcspar fluoresce in a manner comparable with glass.

4. With increasing thickness all bodies become less transparent. In order to find a law connecting transparency with thickness, I made some photographic observations, the photographic plate being partly covered with an increasing number of sheets of tinfoil. Photometric measurements will be undertaken when I am in possession of a suitable photometer.

5. Sheets of platinum, lead, zinc and aluminum were rolled until they appeared to be of almost equal transparency. The following table gives the thicknesses in millimeters, the thicknesses relative to the platinum sheet, and the density:

Thickness.	Relative Thickness.	Density.
Pt. 0·018	1	21·5
Pb. 0·05	3	11·3
Zn. 0·10	6	7·1
Al. 3·5	200	2·6

It is to be observed in connection with these figures that although the product of the thickness into the density may be the same, it does not in any way follow that the transparency of the different metals is the same. The transparency increases at a greater rate than this product decreases.

6. The fluorescence of barium-platino-cyanide is not the only recognizable phenomenon due to X rays. It may be observed, first of all, that other bodies fluoresce—for example, phosphorus, calcium compounds, uranium glass, ordinary glass, calcspar, rock salt, etc.

Of especial interest in many ways is the fact that photographic dry plates show themselves susceptible to X Rays. We are thus in a position to corroborate many phenomena in which mistakes are easy, and I have, whenever possible, controlled each important ocular observation on fluorescence by means of photography. Owing to the property possessed by the rays of passing almost without any absorption through thin sheets of wood, paper or tinfoil, we can take the impressions on a photographic plate inside the camera or paper cover whilst in a well lit room. In former days this property of the ray only showed itself in the necessity under which we lay of not keeping undeveloped plates wrapped in the usual paper and board, for any length of time, in the vicinity of discharge tubes. It is still open to question whether the chemical effect on the silver salts of photographic plates is exercised directly by the X Rays. It is possible that this effect is due to the fluorescent light which, as mentioned above, may be generated on the glass plate or perhaps on the layer of gelatin. "Films" may be used just as well as glass plates."

Prof. Wright, of Yale University, has obtained a photograph of several objects composed of various substances, an illustration of which we have reproduced from the *Electrical World*. See Figure 22.

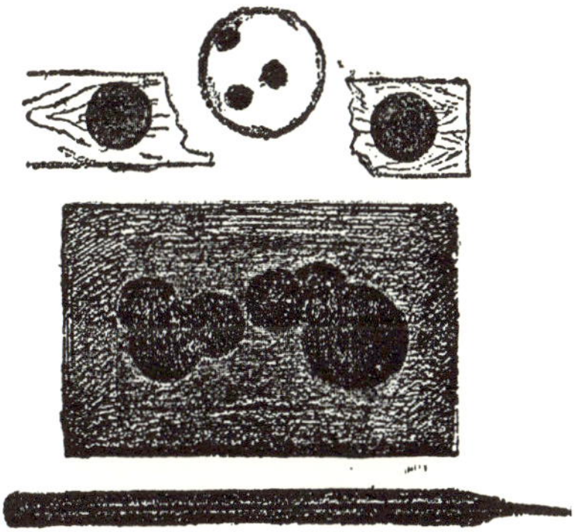

FIGURE 22.

Prof. Wright experiments with a great variety of substances and he found that strong impressions were obtained upon a photographic plate even when it was enclosed in a wrapping of black paper and covered with a pine board half an inch in thickness.

Pieces of glass were found to be more opaque to these rays than some metals or than ebonite, which is perfectly opaque to luminous rays. Among the metals, aluminum is especially distinguished as transparent. In one of the experiments of Prof. Wright an aluminum medal left an impression upon the plate so clearly as to show both design and lettering. In this latter case, the layer between the medal and sensitive plate was an absolutely opaque sheet of ebonite, such as is used by photographers for plate-holder slides.

In other experiments made by Prof. Wright, a closed paper box containing aluminum grain weights was shadowed upon the plate, the appearance being as though the box were almost transparent and the weights themselves somewhat translucent. An ordinary lead pencil lying near the box showed its graphite core by a darker tracing through the middle of the wood of the pencil. A paper box containing, imbedded in cotton, three spheres, one of platinum, one of brass and the other of aluminum, was shadowed, and the box and the cotton were so transparent as to leave but slight impressions on the plate. The brass and platinum spheres intercepted a large portion of the cathode rays, the aluminum a much smaller proportion. A number of American coins, silver, copper and nickel, left strong impressions, showing an almost complete interception of the rays. The copper coins transmitted more than the nickel and the nickel more than the silver. See Figure 22.

Another experiment of Prof. Wright, which

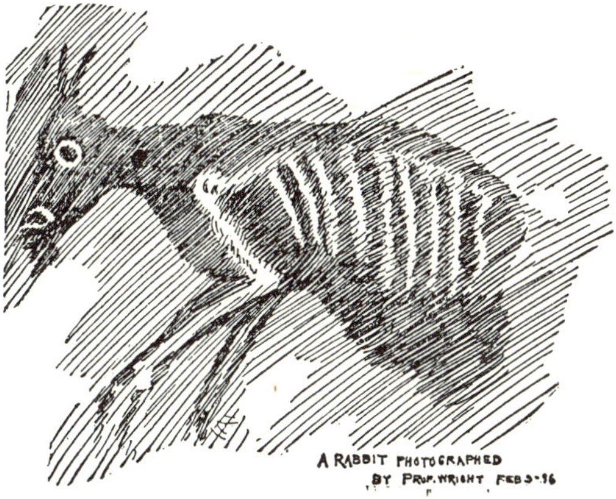

FIGURE 23.

we reproduce from the *Boston Post*, is a photograph of a rabbit. See Figure 23. He says:

"The experiment with the rabbit was not wholly successful, because I happened to have a bad plate; but it shows what can be done. The interior of that dead rabbit was laid bare apparently. This rabbit was covered with heavy fur, and it seems impossible that light could go through him. But it did. And his ribs, the bones in his legs and his backbone are visible in the negative."

The New York World's Experiments.—The *World* recently engaged the services of a

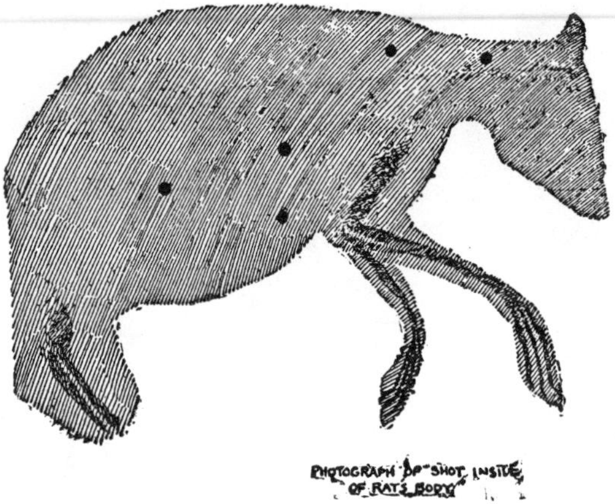

FIGURE 24.

corps of competent assistants, under Prof. Max Osterberg, a fellow of Columbia College and a pupil of the celebrated Prof. Michael I. Pupin. These experiments were made in Prof. Pupin's laboratory in Columbia University for the sole benefit of the readers of The *World*.

Prof. Osterberg succeeded in obtaining a photograph of a pair of scissors, a knife, a cigar-clipper, two nails and a hairpin, all of which had been placed inside an aluminum case. He also took a picture of a comb inside a leather case and a pair of eyeglasses in a

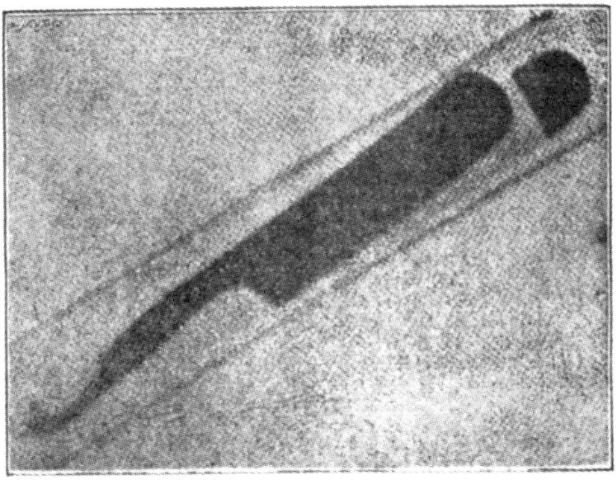

FIGURE 25.

leather case. He succeeded in photographing the backbone of a flounder through its skin and flesh, and had almost succeeded in capturing the skeleton of a mouse, when a remarkable occurrence broke up the experiment.

Figure 24 is reproduced from the New York *World* and illustrates a rat with shot in his body which you can plainly see in the photograph.

Figure 25 is that of a horn-handled razor in its case of cloth-covered cardboard. This picture was taken by Mr. Swinton and was originally published in the *Electrical Review* of London, England.

Figure 26 shows a shadow picture of ten different objects which were laid on the outside of the plateholder and exposed to the X Rays, 30 seconds.

The objects are as follows:

Number 1, Sheet Brass.

Number 2, Aluminum.

Number 3, Diamond Ring.

Number 4, Eye Glass.

Number 5, Mica.

Number 6, Tin Foil.

Number 7, Copper.

Number 8, Steel Key.

Number 9, Hard Rubber.

Number 10, Fine Wire.

Number 11, Silver Ten Cent Piece.

Time of exposure, 30 seconds.

It will be seen by the illustration that aluminum is the only metal that is penetrated by the X Rays to any great extent. A very few get through glass. Hard rubber and aluminum are penetrated about the same, while mica is almost

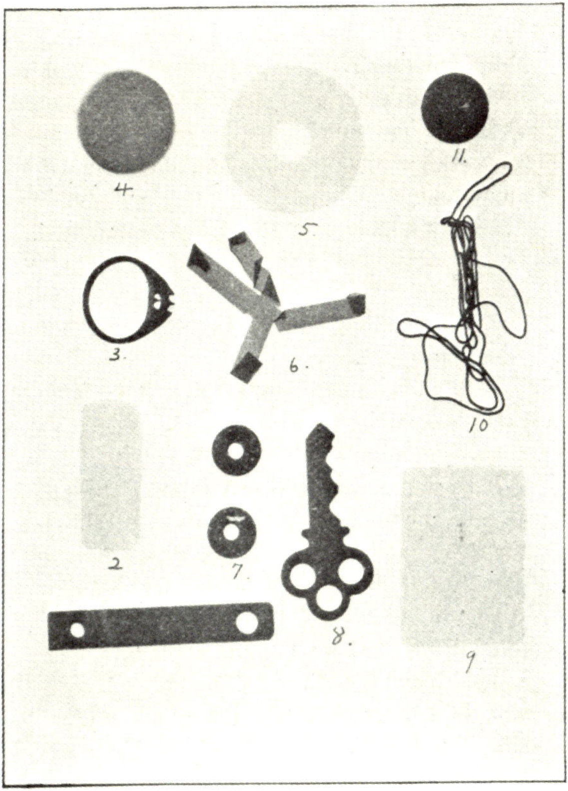

FIGURE 26.

transparent to them. Figure 27 shows the bones of the hand with a diamond ring on the third finger. It will be noticed that the bone is distinctly visible through the diamond. A loss of bone in the fourth finger is also shown caused by amputation. Time of exposure three minutes.

Figure 28 shows a foot of a woman with a needle embedded in the point of the second toe. The needle was located by this shadow picture and afterwards removed by the knife at the Salem hospital.

Fig. 29 is an excellent picture of the bones of the hand, wrist and part of the arm. Two rings are shown on the third finger.

If the reader will examine this picture closely, he will notice with what distinctness the spongy ends of the finger bones are shown. Time of exposure ten minutes.

Figure 30 shows two hands taken at the same time, and on the same plate. A shows a lead pencil with a nickel plated tip. It was laid purposely on the plate holder beside the hands. The hand on the right is that of the author. Time of exposure, seven minutes. All of the foregoing pictures were taken in daylight, no darkening of the room is necessary.

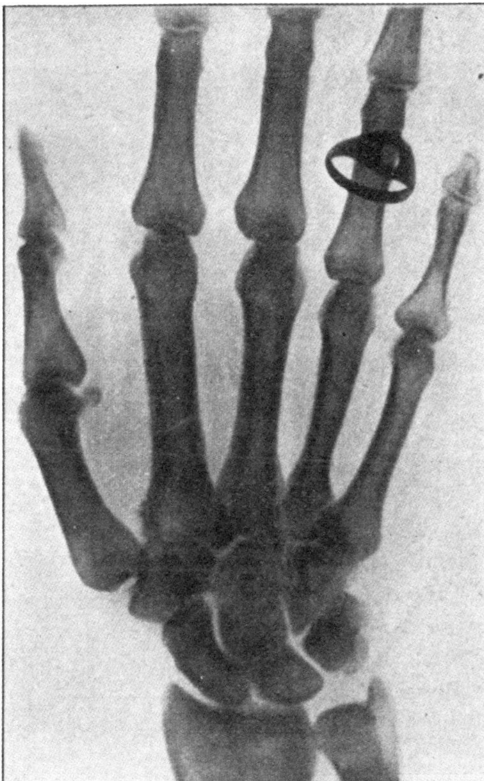

FIGURE 27.

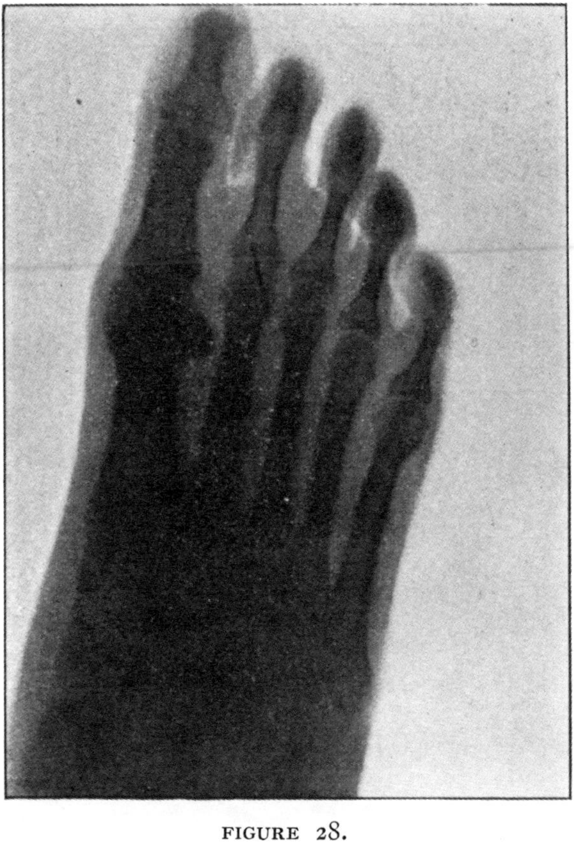

FIGURE 28.

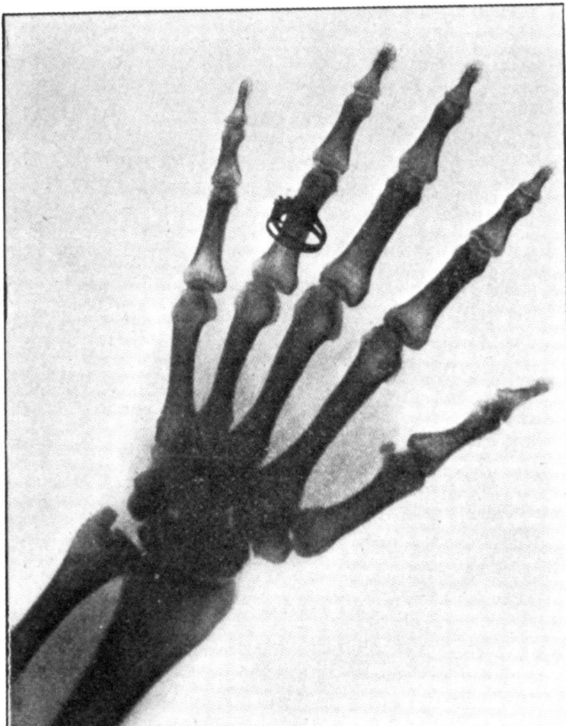

FIGURE 29

Most of the pictures for this book were taken by Mr. George Newcomb of Salem, Mass., who has fitted up a laboratory especially for this business, and is now doing all the City Hospital work. The author takes pleasure in recommending both him and his work to anybody who may require his services.

Figure 31 is a shadow picture, showing a neglected dislocation and fracture of the shoulder joint of a Lynn boy, which resulted in astenomer, (bone tumor). The enlarged bone may be seen by following the shadow A around the shoulder joint. This was taken through the clothing and bandages. The two shadows B and C are the safety pins which held the bandages in place. The diagnosis was confirmed by this shadow picture; the arm was amputated; time of exposure unknown to the author; taken at the General Electric Co.'s factory at Lynn, Mass.

Mr. J. S. Stokes and Mr. M. E. Leeds, of Queen & Co., Philadelphia, have taken a number of successful shadow pictures with Crookes tubes made by their firm. One of these we illustrate in Figure 32, being a picture of a pocket book containing two coins and a key, which, as will be seen, are extremely sharp in outline.

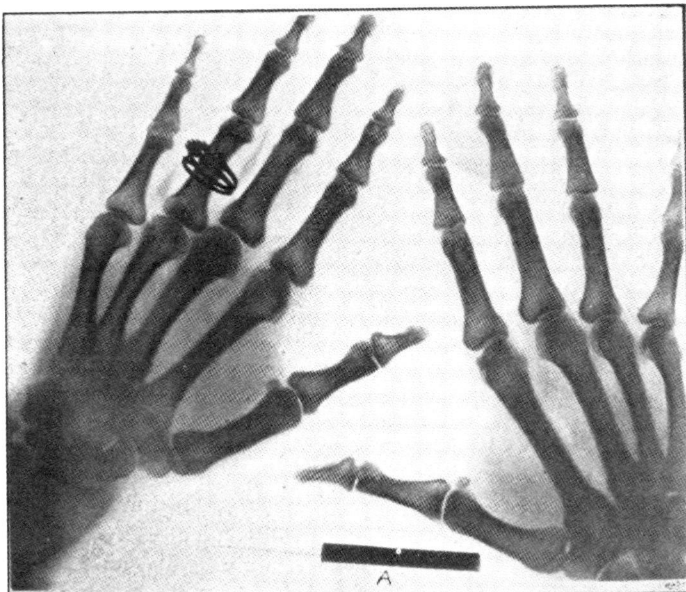

FIGURE 30.

This illustration was kindly loaned by the *Electrical World*, New York.

These shadow pictures were furnished particularly for this book, and the publishers caution everybody not to use or copy them.

*We reproduce from the *Illustrirte Zeitung* two very beautiful examples of X ray photography. The hand is of special interest as being the first photograph (see Fig. 34) that we have seen that shows clearly the position of the veins in the hand. The effect was produced by injecting a fluid in the hand of a corpse, thus making the veins opaque to X Rays and enabling them to be photographed.

Among the experimenters who have lately had remarkable success in photography by means of the Röntgen rays is Dr. Fritz Giesel, of Braunschweig, whose photographs in natural colors have already made a reputation for him in the photographic world. His pictures of still life, showing the natural colors of fruit, flowers and birds, may certainly be classed as some of the best work of this kind ever done, and his success with the puzzling Röntgen rays is proved by the accompanying reproduction of photograph of a canary bird taken immediately after death (Fig. 33). The rays passed unimpeded

*Furnished to us by the Scientific American.

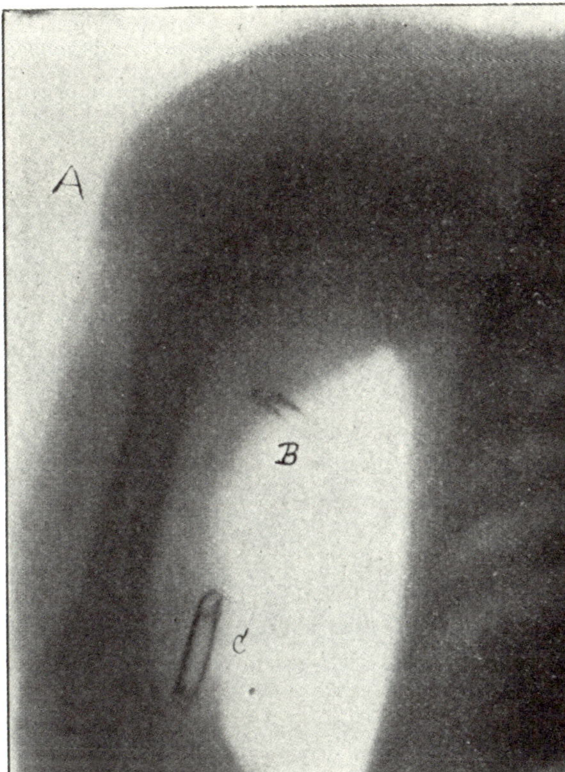

FIGURE 31.

FIGURE 32.

through the feathers, nothing of which shows, the fleshy parts are lightly outlined, but the impenetrable bones have come out distinctly.

Our second illustration Figure 34 is the photograph of the hand of a corpse, taken by means of the Röntgen rays, by Mr. Haschek and Dr. Lindenthal, in Prof. Franz Exner's physico-chemical institute in Vienna. To them belongs the honor of being the first to apply the wonderful discovery of the Wurzburg investigator to a new branch of research. The veins, etc., in the hand—which was the hand of an old woman—are shown by the injection of Teichmann's mixture, which consists of lime, cinnabar and petroleum.

Figure 35 shows the bones of a man's leg, with a fracture of the tibia (the large bone). It was improperly set and caused the fibula (the small bone) to bow out. An operation became necessary and a portion of the fibula was removed by another surgeon, the leg splinted, and the bone allowed to grow together again. A shows the poor work and B the good work. It will be seen by this illustration that an incompetent or careless surgeon's poor work, can be exposed by X Rays.

Figure 36 shows a shadow picture of the arm of a Lynn man, Mr. K. a, veteran of the late

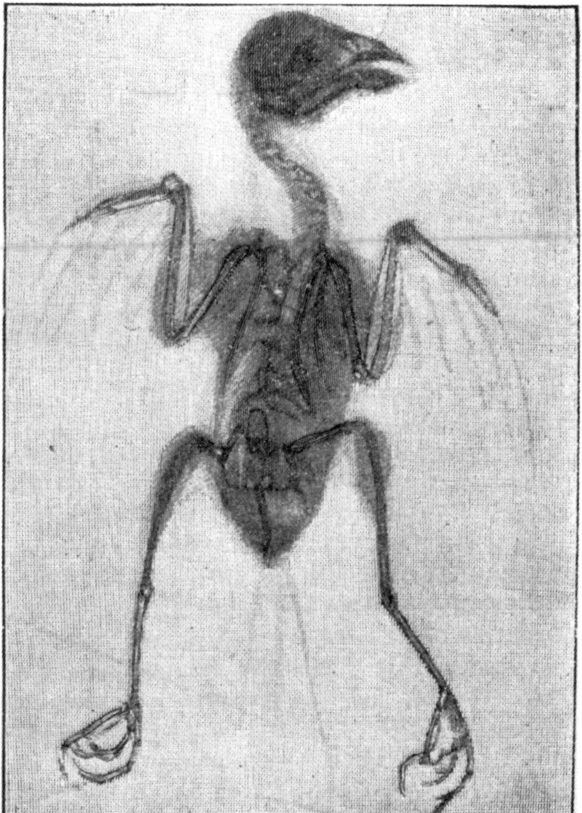

FIGURE 33.

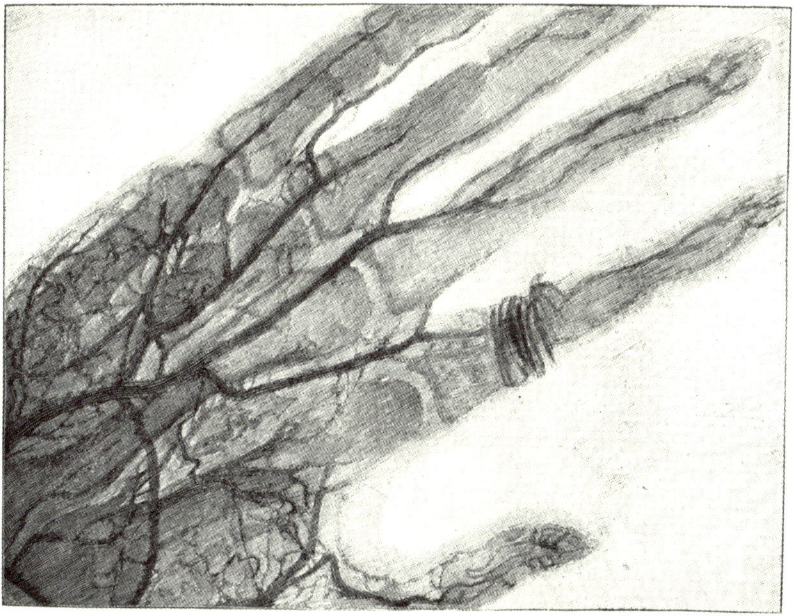

FIGURE 34.

civil war. He enlisted in the 17th Massachusetts Regiment. In the battle of Goldsboro, Dec. 17, 1862, he received a wound in the left arm, the bullet penetrating the forearm about five inches from the wrist. He took part in two other engagements afterwards, and it was nine days before the wound received medical treatment. Physicians probed for the ball, but could not find it, and after two months the wound healed, and from that day to this Mr. K. believed that the bullet was lodged in his arm, though it never gave him any trouble.

The writer became acquainted with this fact and invited Mr. K. to have an X Ray shadow picture made in view of locating the ball. The tube was arranged so as to make a shadow of the arm from the elbow down, ten inches towards the wrist, the part where the bullet was supposed to be embedded. As Mr. K. possesses a large arm a long exposure was deemed necessary to make the experiment a success, but he became somewhat impatient and the rays were allowed to penetrate the member 25 minutes. Fifteen minutes were required in the developing process, and when the plate was brought in for examination, the bones were perfectly shown, but not the slightest traces of a bullet could be found.

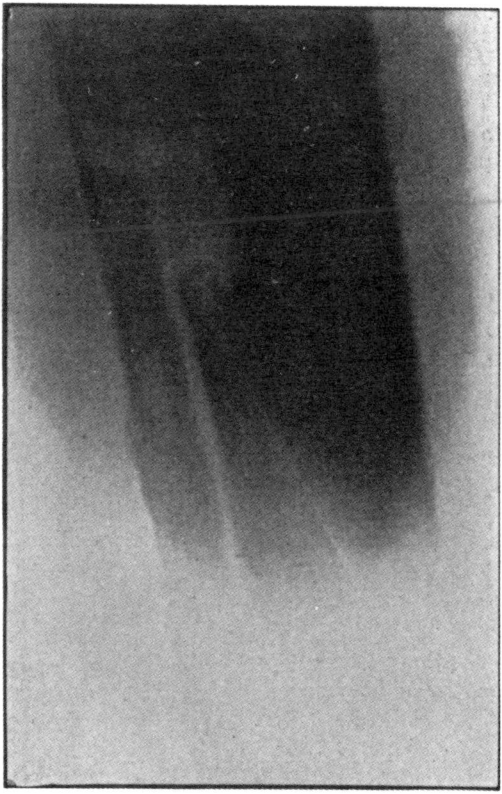

FIGURE 35.

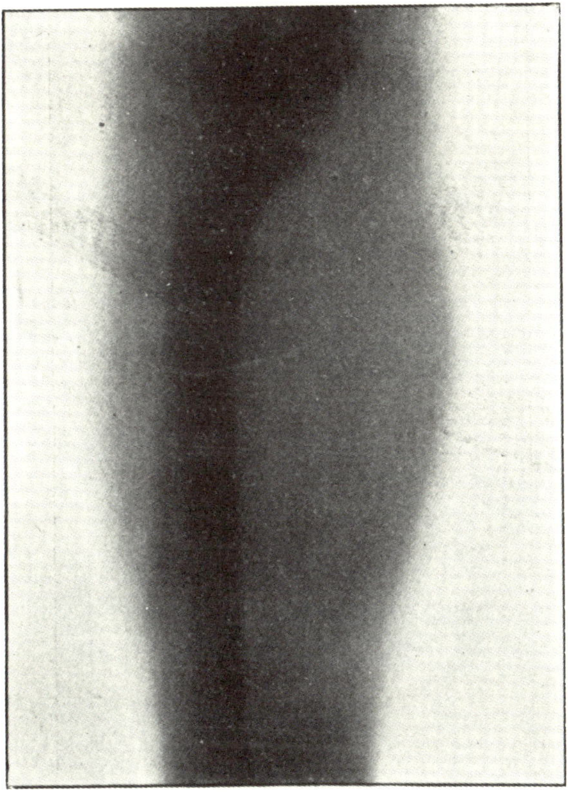

FIGURE 36.

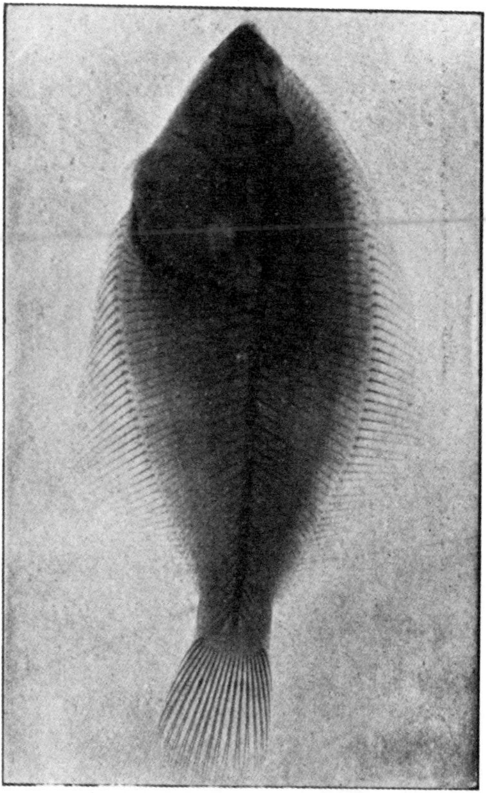

FIGURE 37.

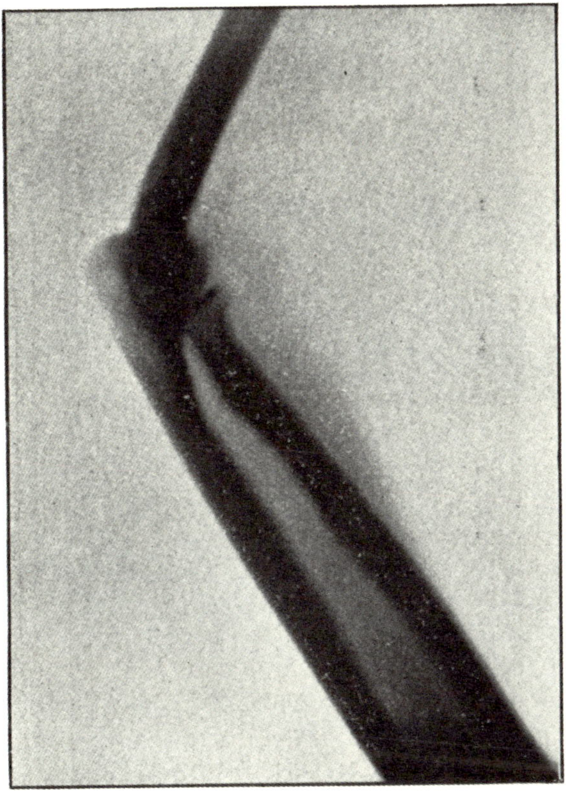

FIGURE 38.

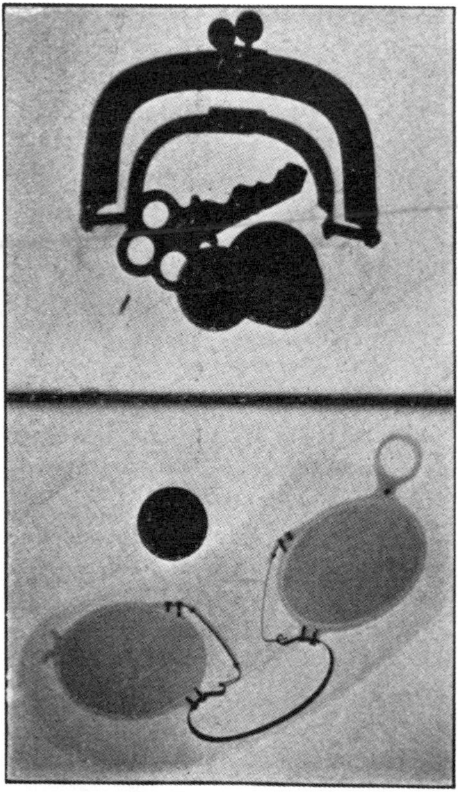

FIGURE 39.

As the doctors and nurses at the time told Mr. K. that the ball could not be found and was not removed, he was greatly surprised at the result of the experiment.

Figure 37 shows a shadow picture of a flat-fish exposed only three minutes. Pieces of shells and other foreign substances in the stomach are plainly visible, as are also the bones.

Figure 38 shows the bones of a young woman's arm, consisting of the humorus (upper bone,) the hinge joint of the elbow; and the ulna and radius (the two lower bones). A fracture of the radius near the elbow is distinctly shown. Time of exposure ten minutes.

In many of the foregoing pictures the articulation of the joints is clearly shown.

Figure 39 shows a pocket book containing coins, a key, and a bit of brass, also a pair of eyeglasses contained inside of a leather case. Time of exposure three seconds. This is one of the finest shadow pictures we have ever seen. In the negative the threads of the screws inside the hard rubber frame of the glasses can be distinctly seen. Taken for this book by Mr. Newcomb.

CHAPTER III.

The Fluorscope and other Apparatus.

The fluorscope is a device to be used in visual examinations of opaque objects by means of X Rays. It consists of a wooden box made light tight, and shaped as seen in the engraving.

On the end to be held before the Crookes tube is fastened a sheet of cardboard, one side is covered with a crystalline chemical salts which become fluorescent when placed in the path of the X Rays.

By placing an opaque body between the tube and the cardboard of the fluorscope a shadow is thrown upon it and is visible to the human eye. It was invented by Edison who discovered that platino-cynanide of barium had the property of fluorescence and was the salts originally used. He has since substituted the tungstate of calcium which he found has the same properties to a much higher degree. The crystals are very minute and are distributed very evenly over the cardboard.

Any object may be seen instantly by the fluorscope if the X Rays are being developed.

FIGURE 40.

Fig 40 shows the fluorscope in use, the experimenter is making an examination of his hand. To use this device most successfully you must have a good strong current and an induction coil giving not less than a spark of four inches.

An X Ray Meter.

The peculiar glow exhibited by a "focus" tube working well furnishes a good criterion of efficiency as regards Röntgen rays. A more definite means of comparing the actinic power of the radiation has been produced by Messrs. Reynolds and Branson, Leeds. A small quadrant of aluminum is constructed in concentric terraces, ranging from one millimetre to ten millimetres in thickness. By holding this quadrant between an excited Crookes tube and a phosphorescent screen, the thickness of aluminum which the rays are capable of traversing can be seen upon the screen; or, by substituting a sensitive plate for the screen, the effect may be photographed. The "X Ray meter," as the quadrant is called, thus furnishes an easy means of comparing the intensity of Röntgen rays emitted by different tubes and by the same tubes at different times.—*Electrical Engineer.*

New Facts About X Rays.

Dr. Röntgen has published some new facts about his rays. He finds that all solid bodies can generate them; the only difference being in the intensity, the greatest intensity being produced by platinum. He finds that the insertion of a Tesla coil between the Ruhmkorff coil and the ray-producing apparatus is very advantageous, and that the X Rays and the air traversed by them can discharge electric bodies. —*Scientific American.*

On the Action of the X Rays Upon the Diamond.

The transparency of the different varieties of carbon and of its non-metallic compounds, established by Professor Röntgen and then by the experimentalists who have studied the X rays, may serve to distinguish clearly the diamond from its imitations made of very opaque substances.

The proofs which we have the honor to submit to the Academy show in juxtaposition silhouettes of genuine diamonds and of imitations both loose and set. Prolonged exposure soon succeeds in causing the silhouettes of genuine diamonds to disappear, while false diamonds continue to behave like opaque substances. The same procedure has also allowed us to distinguish natural jet from its mineral imitation.

In addition to this graphic method we have tried an optical method, in which we have tried the fluorescence studied by Professor Röntgen. Diamond and jet, if interposed between the Crookes tube and a leaf of paper covered with a fluorescent substance (*e. g.*, barium platino-cyanide), project upon the paper shadows lighter than those which appear beneath imitations placed near.

Here then we have two very certain tests: the *graphic* method leaves an irrefutable document, while the *optical* method is instantaneous. They will easily come into practical use, since a precious stone may be tested even in its setting, and without running any risk.—*Comptes Rendus*, CXXII., p. 457.

How to Make Tungstate Calcium.

Mix about 1 ounce each of common salt, tungstate soda, and choride calcium; last two articles should be bought at retail for about 2 cents per ounce. Put the mixture in a common crucible, also obtainable for about 10 cents, fit a tin cover to it and bury to the lid in a good coal fire—the kitchen stove will do—so as to bring it to a full red heat; leave it for two or three hours, or until contents are fused to a clear liquid, then set it out to cool and crystallize. The resulting hard, glass-like mass should be broken out with an old chisel or by breaking the crucible—broken up and thrown into a jar of water, which will gradually dissolve the chloride of sodium formed, and the fine crystals of tungstate calcium will settle to bottom. Wash by decantation till all taste of salt is gone, and pour out on filter or blotting paper and dry.—H. G. OGDEN in *Scientific American.*

Diffusion and Opalescence with Roentgen Rays.

So far as the writer is aware, the phenomena presently to be described have not hitherto been alluded to by writers or experimenters on Röntgen rays. They are the phenomena of diffusion of such rays by certain classes of substances in such a way that such substances must come to be regarded as opalescent,* or to act like opal glass in ordinary light, or like milky liquids, a diffusion of incident rays taking place from the interior and exterior of the mass.

Some substances are found to behave with Röntgen rays in the way that compacted snow or translucent ice act in diffusing light.

---

*Opalescence is here used to indicate the property of milkiness without play of color.

Let a fairly large metal screen, such as a brass or iron plate, of one-sixteenth to one-eighth of an inch in thickness be suspended in a vertical plane about a foot or more from a Crookes tube kept in action and emitting rays toward the plate. On the side of the plate opposite to that nearest the Crookes tube there will, of course, be a space in shadow free from Röntgen rays. This statement, though true in the abstract, may not express the whole truth. It may easily happen that an exploration with a fluorescent screen tube, or fluoroscope, back of the metal plate will show that the rays are not absent there, but appear to come around the edge of the plate. This phenomena has been recently alluded to by Mr. Edison, and used in support of his suggestion that Röntgen rays are simply high-pitch sound waves.

But upon close examination it will be found that the rays back of the metal shield are at least mostly due to diffusion from surrounding objects; the walls of the room, or the floor, or objects in the room receiving the rays and scattering them. Further examination will also disclose the fact that bodies differ greatly in their diffusive power, and that substances such as paraffin, wood, paper, pure rubber, cloth of cotton or wool, the hand of a person, etc., which are comparatively transparent to the rays, are also fairly vigorous in diffusing them.

The diffusion is not merely from the surface or irregular reflection. This is shown by the fact that the surface may be smooth or rough, and that the rays come even from the back of the object and from portions not exposed directly to the original rays. This indicates a true opalescence like that possessed by opal glass with ordinary light. It will be seen from this that the shadows of opaque objects embedded in tissue at considerable depths can never be so black or dense as when the objects are merely surrounded by air.

Liquids appear to possess the property as well as solids. The diffusion appears to take place in all directions within and outside of the mass of the substance.

If the fluorescent screen tube used has metal sides of some thickness, the large metal plate screen mentioned above can be dispensed with. In this case it suffices that the screen tube be turned so as to be directed at right angles to the direction of the rays or be turned away from the Crookes tube, so as not to fluoresce. Pieces of various substances may now be placed opposite the end of the screen tube, but in a position to be partially or wholly exposed to the rays. The pieces will become virtually sources of the rays, and the rays diffused by them reaching the fluorescent screen will cause it to emit light.

By placing two exactly similar fluorescent screens at opposite ends of a tube and employing a Bunsen photometer screen movable, as usual, between the screens, a comparison of diffusing power of different materials may be made by subjecting the pieces placed near the ends of the photometer tube outside, to equal radiation from the Crookes tube.

The same instrument might be employed to measure the relative merits of the Crookes tubes as producers of the rays. The same electrical discharges could be passed through both tubes while opposite each end of the "fluorometer" so also the comparative values of different fluorescent screen materials could be tested by slight modifications in the use of the instrument.

By comparing the density of the shadow of a piece of sheet lead, when back of a block of wood or paraffin and close to the fluorescent screen, with that of the same piece when the wood or paraffin is between the screen and the lead, the effects of diffusion in the latter case in lighting up the shadow are clearly seen on the fluorescent screen.

Though the experiment has not been tried as

yet, the writer thinks that a mass or cloud of fine particles of water from a steam jet would diffuse the Röntgen rays. It is well known that the higher pitch light waves are most readily diffused and absorbed by fog, and if Röntgen rays be very high-pitch waves similar to light waves this diffusibility need not surprise us; the new rays to be in a sense detectors of molecular turbidity.

Whether these gases have the power of diffusing the rays is not known, but the diffusive power, if it exists at all, will be small and difficult to detect.

Metal plates gave apparently little diffusive effect, appearing to reflect feebly at angles equal to the incidence angles. The phenomena of reflection from surfaces have been investigated very recently by Mr. Tesla and a number of values obtained for the relative reflecting powers of metals, etc.

The diffusive action herein noted is different altogether from reflection, and is obtained at all angles from the surfaces, and from the interior of the materials upon which the original rays fall.—*Electrical World.*

*Edison's New X Ray Lamp.

A notable example of the stimulation of invention by new discoveries is found in the latest work of Edison, which follows the discovery of Röntgen and the fluorscope of his own invention. This latest invention is a fluorescing lamp in which is found the promise of the artificial light of the future. The lamp appears to have all the qualities requisite for perfect illumination; the light is mild but effective; it is diffusive like daylight. It gives off no perceptible heat, which latter duality goes to show that its economy has no parallel in other kinds of artificial illumination.

One form of the lamp consists of a highly exhausted oblong glass bulb having wires sealed in the ends, each wire being provided with a small plate inside the bulb, one of these plates being inclined to cause a distribution of the rays over the side of the lamp. The inner surface of the lamp is covered with a granular mineral substance which is fused on the glass and is highly fluorescent. When the lamp is excited by connection with an induction coil, the fluorescent material becomes luminous. See Fig. 41.

---

*Illustration and MSS kindly furnished us by the Scientific American.

Originally Edison used calcium tungstate for his fluorescent material; but by trial he found that the vacuum soon deteriorated, and after a long series of experiments has discovered a fluorescing material which does not affect the vacuum, while it has a higher efficiency than the calcium tungstate.

Mr. Edison thinks that the fluorescing material converts all of the X Rays into light. He has a theory as to the manner in which the light is produced. The crystals are composed of light and heavy particles and the impact of the waves produces a stress in the crystals which causes the emission of light. Mr. Edison describes these waves as sound waves, because they differ in their mode of vibration from ether waves. Their motion is infinitely more rapid than that of sound waves with which we are familiar; they are comparable as regards velocity with electric or light waves. As to efficiency, the fluorescing lamp produces light at the rate of 0.3 of a watt per candle power. When this is compared with 3 watts per candle power for incandescent lamps, and ½ watt per candle power for arc lamps, it will be seen that there must be great economy in the fluorescing lamp.

Development of The Negative, etc.

The negative of an X Ray picture is developed in the same manner as those taken by the camera. The writer prefers the Hydochinon. Mr. Newcomb uses the Pyro developing solution. A combination of Eikonogen and Hydrochinon makes a very good developer. Fix the negative with solution of Hypo-Sulphite of

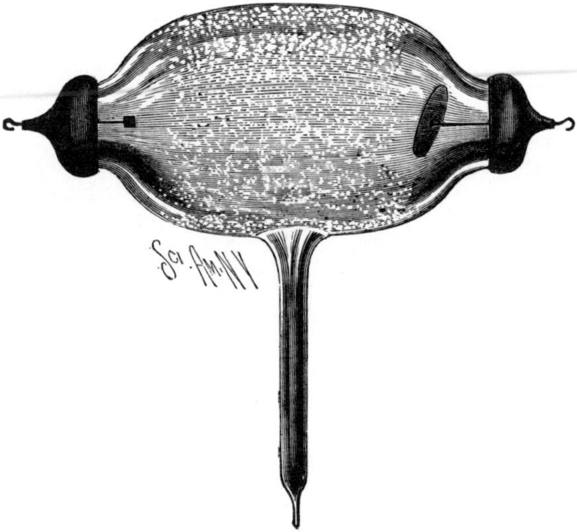

FIGURE 41.

Soda. Printing, toning and fixing the print may be done on any good albumen or gelatine paper the same as in ordinary photography. Ilo paper is exceptionally good, using the Ilo combined toning and fixing solution, or the combined toning and fixing solution made for the Aristotype paper.

The Ruhmkorff Coil.

This is a kind of induction coil invented by Ruhmkorff. Originally the Ruhmkorff coils were made in comparatively small sizes, as it was found impossible to insulate the secondary wire sufficiently to resist the high potential of the electricity. The different layers of wire were laid the entire length of the coil, but sep-

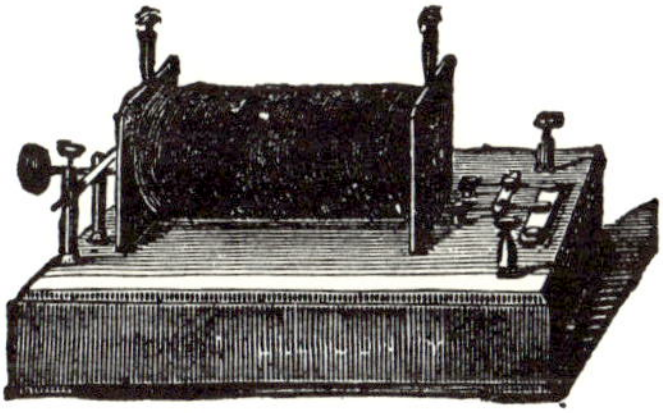

FIGURE 42.

arated from each other by silk. Ritchie wound the secondary in sections, each on a spool of the finished diameter of the cell, but very short, requiring eight or ten such spools in a row to give the requisite total length. These spools were made of hard rubber, and could stand very high potentials, Ruhmkorff bought one of these coils, took it to pieces, and copied the construction in the large sizes of his manufacture. See Figure 42.

INDEX.

A

C

D

E

S

T

W

X

APPENDIX.

Some Various Types of X-Ray Tubes.

A large number of tubes have already been employed in different experiments with, and applications of, the X-rays for photography, and in connection with the fluoroscope. Mr. G. Séguy has constructed and experimented upon several types, and he has gathered a collection which is illustrated in "La Nature."

There exist at the present time three methods of obtaining the X-rays. That employed in the very beginning is based on the direct action of the ray. The second permits of obtaining instantaneity in the radiograph, and is based on a reflection action. The third is a result of the combination of the first two methods.

In the accompanying engravings, Nos. 1, 2, 3, 4, 6, 7, 10, 11, 12, 13, 14, 17, 18, 20, 21, 24, 26, 28 and 32 are constructed according to the principles of the first methods. Nos. 5, 8, 9, 15, 16, 23, 25, 27, 29 and 30 employ the second method; that is, the theory of the reflection of the cathode rays and of the phenomenon of internal electrolysis of the volatilized molecules. The tubes Nos. 19, 22 and 31 produce X-rays according to the two combined theories.

The numbers accompanying each tube designate the design of the various experimenters, as follows: 1 and 2, Crookes; 3, Séguy; 4, Wood; 5, Séguy; 6, Chabaud-Hurmuzescu; 7, Séguy; 8, Thompson; 9, Séguy, 10, d'Arsonval; 11, Séguy; 12, Puluj; 13, Séguy; 14, d'Arsonval; 15, Le Roux; 16, 17 and 18, Séguy; 19, de Rufz; 20, Crookes; 21, 22, 23, Séguy; 24, Röntgen; 25, Brunet-Séguy 26, 27, Le Roux; 28, Colardeau, 29, Séguy; 30, Colardeau; 31, Séguy; 32, Röntgen.

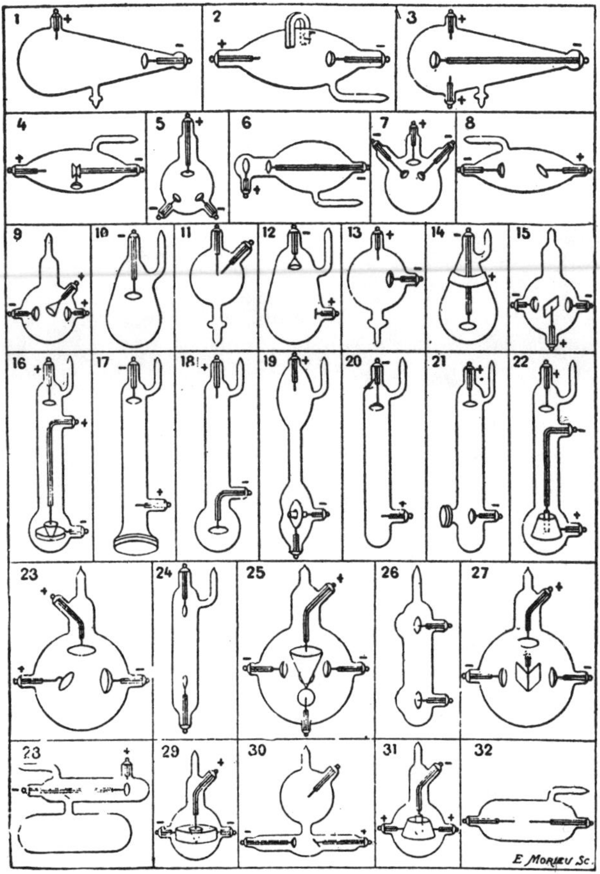

Various Types of X-Ray Tubes.